SO WHAT NOW?

SO WHAT NOW?

Recognizing Six Human Responses to an AI-Driven World

TOM MAWHINNEY

THIN LEAF PRESS

Publication Data
Names: Mawhinney, Tom, Author
Title: *So Now What? Recognizing Six Human Responses to an AI-Driven World*

ISBN: 978-1-968318-56-7 (paperback) | 978-1-968318-55-0 (eBook)

Artificial Intelligence, Psychology, Adaptation, Decision-Making
Thin Leaf Press
Los Angeles

Book Cover Design by 100Covers
Interior Formatting by Dindo Sanguenza.

TABLE OF CONTENTS

INTRODUCTION

I am writing this in 2026, in the middle of a kind of change that is hard to pretend we are not living through. Artificial intelligence is not the only force at work, and it is not the first technology wave we have experienced. It is, however, the one showing up right now in the most visible, disruptive, and strangely personal way. It is not only changing tools. It is changing expectations, conversations, and the quiet assumptions we have been using to plan our lives.

Things are moving quickly enough that I would not be surprised if I am writing a follow-up to this book within the next six to eight months. That is not marketing. It is simply an honest acknowledgment that the ground is still shifting and that a book written today is a snapshot, not a final verdict.

We have all lived through new technologies before. Some arrived with a lot of noise and then faded into the background. Others took longer to prove themselves and then quietly became part of daily life. Most landed somewhere in the middle, useful in some places, irrelevant in others, and less dramatic than the headlines made them seem. What feels different now is not just that AI exists, but that it has moved from "interesting" to "everywhere" faster than most of us expected, and it has done so at the exact moment when many people already feel stretched for time, attention, and confidence.

Not long ago it made sense to treat AI as something to observe from a distance. It showed up in headlines, podcasts, and conference talks. People argued about whether it would replace jobs, supercharge creativity, or turn education upside down. Some insisted it would change everything. Others dismissed it as the latest overblown trend. If we chose to wait and see, that was not laziness or denial. It was caution, and caution has often been the sane response when the internet is in one of its excited phases.

What has changed is how often we run into it now, and how hard it has become to separate the technology from the conversation about the technology.

We see it when our email suggests replies before we finish typing. We see it when a search engine gives a summary instead of sending us down a rabbit hole of links. We see it when our phone offers to rewrite a message in a different tone. We see it when tools promise to create slide decks, images, videos, essays, or code in minutes. We also see it in the less obvious places, like the way ads follow us around the internet, because the marketing machine is now learning just as quickly as the technology itself.

You search once for something like "best resume format," "how to prep for an interview," "how to plan a trip," "how to start a side project," or "how to learn faster," and suddenly AI is everywhere in your feed. You start seeing tools that claim they will write the resume for you, optimize it for recruiters, build the portfolio, coach the interview, and apply for jobs while you sleep. You watch one video about using AI at work and your feed fills with creators showing shortcuts, templates, and "prompts that will change your life." You do not have to be paranoid to notice what is happening. The ads are increasingly tailored. The recommendations are increasingly accurate. The pressure is not only coming from the technology. It is also coming from the attention economy that now surrounds it.

There is another layer to this that we should name directly, because most people feel it even if they do not talk about it out loud. Social media is full of people positioning themselves as experts simply because they moved early. Some of these people are thoughtful and genuinely helpful. Others are enthusiastic, loud, and confident in the way social media rewards. A person can use a tool for two weeks and begin teaching it as if they discovered a new law of physics. If we are being honest, the sheer volume of these voices creates anxiety all on its own, because it turns a technology shift into a status game. It becomes easy to feel late, behind, or irrelevant simply because someone else is posting confidently.

Even if we do not go looking for it, it appears, and it is difficult to stay neutral when something keeps showing up in the spaces where we spend our time.

That steady presence changes how this moment feels. If we are in our twenties trying to build a career, it can feel like the rules are shifting while we are still learning them, and the worry is not just whether we will get a job but whether we are building skills that will still matter by the time we have built a life around them. If we are in the middle of our careers, it can raise a quiet question about how long our current skill set will carry us, especially when we hear talk about automation in roles that used to feel stable. If we have been leading for years, or if we have stepped back from full-time work, the tension often feels different but no less real, because it can feel like the conversation is moving in directions we did not prepare for and that we are expected to have opinions about tools we have not had time to explore deeply. If we are parents, the anxiety can show up as a gap, the sense that our kids are living in a digital world that is moving faster than we can follow, and that we are supposed to guide them even when we are not sure what we are looking at. Even younger users can feel pressure when every scroll brings another creator claiming to have cracked the code, another tool drop, another "you're falling behind if you're not doing this" message.

The anxiety shows up everywhere, and it is not limited to jobs.

It shows up in the way people talk about school and cheating, because the old assumptions about essays and homework are suddenly unstable. It shows up in relationships and communication, because people can now generate messages and responses that sound sincere even when they are not. It shows up in creativity, because a person can generate a hundred images in an afternoon and then wonder what "original" means anymore. It shows up in news and trust, because it is harder than ever to know which sources are reliable, and it is easier than ever for convincing misinformation to spread. It shows up in the background hum of daily life, where many people already feel overloaded and now feel as though they are supposed to learn an entirely new set of tools just to keep their footing.

What makes this moment difficult is not that we lack information. We have too much of it. We have news, reactions, predictions, tutorials, hot takes, and marketing campaigns coming at us constantly, often packaged as certainty. When we are tired, uncertainty feels threatening, and certainty feels comforting even when it is false. That is how people get pulled into extremes, either panic on one side or dismissal on the other. Neither extreme helps in real life.

That recognition is what led me to write this book now rather than later.

In the past, I would have waited for a quieter stretch before committing to a project like this. Writing has always required time and focus, and I have tried to respect that. This time I chose differently. I have used AI tools for a while to edit and refine my professional writing, mostly as a practical way to tighten language and test clarity. With this book, I leaned on them considerably more, and I did it on purpose.

If the message of this book is that AI is best approached as a complement rather than something to worship or fear, then it makes sense to write in a way that reflects that posture. The perspective in these pages is mine. The experiences are mine. The judgment about what belongs and what does not belongs to me. What changed was the way I moved from outline to finished manuscript, because I allowed the tools to help me draft faster, reorganize more easily, and pressure-test ideas quickly enough that the book could be completed in days rather than months.

I did not make that choice because I think speed is everything. I made it because the pace of change has made timing part of the message. In slower cycles, you could wait for the dust to settle, write with perfect hindsight, and still publish something relevant. In this cycle, waiting for perfect clarity can easily become a reason to never engage at all, and by the time you do, the conversation has moved on and the anxiety has deepened.

So yes, the way this book was written is part of the point. It is a living example of what I am suggesting throughout these pages, which is that we do not need to compete with the tools or pretend they are not there. We can learn how to work alongside them in ways that support our goals, whether those goals are professional, personal, or simply emotional.

It is also important to be clear about what this book is and what it is not. It is not a guide to becoming an AI expert, and it is not an attempt to predict which tools will win or lose. Artificial intelligence happens to be the force driving this particular moment, but the underlying principle is broader than any single technology. When the conditions around us change in a meaningful way, the sustainable response is thoughtful adjustment rather than denial or panic. That belief has shaped much of how I think about leadership, strategy, and personal posture, and this book simply applies that same belief to the current wave of AI. The topic may feel different. The logic is the same.

That brings us to the real reason I wrote this book.

Most people are not looking for a technical explanation of AI. They are looking for a way to feel steady again. They want to understand what matters, what does not, and what they can do that will actually help. They want to stop feeling whiplash every time a new tool appears or a new prediction goes viral. They want to know how to respond in a way that reduces anxiety rather than feeds it.

We tend to respond to change like this in a handful of predictable ways. Some of us lean in quickly because we see opportunity, and we want to be early rather than late. Some of us hang back because we do not trust bold claims, and we are wary of being manipulated. Some of us feel exposed, because we sense that parts of the world we understood are shifting and we do not want to be caught pretending we are comfortable when we are not. Some of us feel overwhelmed, because the volume of information is exhausting and the mental load is already high. Some of us feel outdated, because the culture around these tools seems to be moving without us, and we do not know how to step back into the conversation without feeling foolish. Most of us move between these states, depending on the day, the context, and what else is going on in life.

I certainly have.

There were times when I dismissed what I was seeing because it sounded exaggerated, and I did not want to waste time chasing noise. There were stretches when curiosity led me to try more tools than I could realistically use. There were moments when I felt exposed, because I could feel the conversation shifting and I knew I had not spent enough time testing what people were talking about. There were weeks when the constant stream of content left me more tired than informed, and the tiredness made everything feel heavier than it needed to be. There were moments when I wondered whether I was adapting thoughtfully or simply reacting to whatever appeared next.

None of these reactions are irrational. In many ways, they are healthy instincts. Skepticism protects us from hype. Ambition pushes us to learn and improve. Curiosity helps us explore without fear. Staying informed helps us avoid blind spots. Experience helps us keep perspective when others overreact. The difficulty is that when the environment changes, the same instinct can lead to a different outcome than it once did.

Working harder used to be a reliable way to stand out when output was scarce. When a machine can generate a draft in seconds, effort alone does not separate us in the same way, and a person can burn themselves out trying to compete on speed. Ignoring hype used to conserve energy. When tools are quietly shaping how people study, write, build, and communicate, ignoring everything can mean missing something that would have been helpful. Reading more used to create an edge. When information is constant, reading more can simply create noise, and noise is fuel for anxiety.

This is where the personas in this book come in: Ambitious, Exposed, Curious, Skeptic, Overwhelmed, and Outdated. They are not labels meant to define us. They are mirrors, and most of us will recognize ourselves in more than one of them.

Each persona is paired with a possible evolution. Not because change is mandatory, and not because the starting point is wrong, but because the context has shifted enough that a small adjustment can often produce a better outcome. If we are ambitious, the question becomes whether we are chasing output or building the kind of value that still matters when machines can draft quickly. If we feel exposed, the question becomes whether a small increase in familiarity would make us steadier rather than anxious. If we are skeptical, the question becomes whether we are protecting ourselves from hype or simply avoiding firsthand experience. If we are overwhelmed, the question becomes whether we need more information or less, and whether reducing inputs might restore clarity. If we feel outdated, the question becomes whether we need mastery or simply enough understanding to participate and feel connected again.

None of these shifts are imposed. They are available. We can choose to stay where we are if it is serving us, and we can choose to adjust if it is not. That is the posture of this book. It is not a sermon and it is not a blueprint. It is a set of reflections and suggestions offered in the middle of a moment that is still unfolding.

I do not know exactly what this landscape will look like even a year from now. I am writing in the middle of it, aware that some of what feels urgent today may look obvious later and some of what feels clear today may require revision. That uncertainty is not a weakness.

It is simply honesty, and it is also part of the reason a follow-up edition may be needed sooner than in a normal publishing cycle.

What does feel clear is that stepping back entirely carries a cost, even if that cost is slow and quiet rather than dramatic. Over time, it narrows options. It narrows confidence. It narrows participation. Engaging, even in modest ways, tends to reduce anxiety because experience replaces speculation. Trying a tool ourselves often clarifies more than reading ten opinions about it, and using it in our own lives reveals its strengths and limitations far better than any viral post.

This book is my attempt to translate that lived experience into practical reflection, written in plain language for people who have better things to do than chase every tool release and every headline. It is not a technical manual. It is not a forecast. It is an invitation to respond deliberately rather than reactively, and to decide how we want to position ourselves alongside tools that are clearly not going away.

The title asks a simple question: So now what?

It is the question that follows the realization that something meaningful has shifted, and it is also the question that moves us from watching to deciding.

The tools are here, the noise is loud, and the pressure is real.

What remains open is how we choose to meet it.

Earlier I mentioned a handful of postures that tend to surface when change accelerates the way it is accelerating now. Those personas are not theoretical. They are patterns most of us recognize in ourselves when the ground shifts. The chapters that follow take a closer look at them, not to box anyone in, but to make the starting point clearer. Before we can decide what to do next, it helps to understand the posture we are bringing into the moment.

PERSONA #1: THE OVERWHELMED

When change accelerates the way it is accelerating now, one of the most common responses is not excitement or ambition. It is fatigue.

If you have found yourself feeling mentally crowded every time you open your phone or sit down at your desk, this may sound familiar. You scroll and see another post about how everything is changing. Another new tool. Another update. Another confident voice explaining how to automate half your life. You read an article about jobs disappearing or being "reinvented." You watch a short video that makes something look effortless and think you should probably look into it. Then you close the tab because you are already tired.

Overwhelmed does not usually arrive as panic. It arrives as saturation.

It looks like bookmarking articles you never return to. It looks like downloading a tool, clicking around for a few minutes, and deciding you will come back to it when you have more time. It can also look like telling yourself you are "aware" of what is happening while quietly avoiding deeper engagement, or knowing just enough to feel unsettled but not enough to feel steady.

Overwhelmed often shows up in small, almost forgettable moments. You open a shared document at work and realize a colleague clearly used AI to draft it, and you feel a brief flash of discomfort because you are not sure how they did it. You sit in a meeting where someone references a tool you have heard about but never tried, and you nod along while quietly promising yourself you will look into it later. At home, your child mentions using AI to outline a paper, and you realize you do not quite know what that means in practice. None of these moments are dramatic. Taken together, they create a steady hum of unease.

This response makes sense.

We are living in an environment where information does not just arrive; it stacks. AI is not only changing tools. It is changing the rate at which new claims, updates, and opinions are produced. There is always something new to learn, test, or worry about. For someone already balancing work, family, and ordinary responsibilities, the idea of adding "figure out AI" to the list can feel unrealistic.

There is also a psychological layer to this. When something feels large and uncertain, the brain looks for relief. One form of relief is narrowing attention. We tell ourselves that if we ignore the noise, we can preserve energy. In earlier cycles, that instinct often worked. Not every platform or product deserved our time. Waiting filtered out a great deal of hype.

The difference now is that the signal and the noise are mixed together. Some of what looks like hype becomes embedded in everyday tools. Some of what feels optional quietly becomes expected. The line between "this will pass" and "this is staying" is harder to see in real time.

When overwhelmed becomes the dominant posture, two patterns tend to follow.

First, avoidance becomes habit. The longer we delay engaging with something that feels complex, the larger it appears in our minds. What might have taken thirty minutes to explore six months ago now feels like a gap that requires hours to close. The psychological barrier grows faster than the practical one.

Second, we start to confuse exposure with understanding. Because AI appears in our feeds and conversations constantly, we assume we are keeping up. In reality, passive exposure often increases anxiety without increasing competence. We know enough to sense that change is happening, but not enough to feel confident inside it.

Over time, staying in a constant state of overwhelm has a quiet cost. It narrows confidence. It makes experimentation less likely. It widens the gap between what others are trying and what we understand firsthand. It does not collapse our world overnight. It creates slow drift. The drift shows up in smaller ways at first. We hesitate to apply for a role because it mentions AI familiarity. We avoid suggesting new ideas because we are not sure how tools might factor in. We opt out of conversations rather than risk sounding uninformed. None of these choices feel catastrophic. Together, they reduce agency.

None of this means that feeling overwhelmed is weakness. It is a rational response to sustained stimulus. The question is not whether the overwhelm is justified. The question is whether staying there serves you.

If you recognize yourself in this posture, the evolution is not toward becoming hyper informed or endlessly productive. It is toward becoming Focused.

Focused does not mean consuming more information. It means choosing deliberately where to engage. Focused people do not try to understand everything. They decide what deserves their attention and allow the rest to pass without guilt.

Focused might look like deciding that for the next month you will use AI only to help with one recurring task at work, and ignoring everything else. It might mean asking your child to show you how they are using a tool instead of pretending you already understand it. It might mean unsubscribing from five AI newsletters and keeping one. It might mean choosing two tools to explore and deliberately ignoring the constant stream of new releases.

The shift from being overwhelmed to becoming Focused is less about intensity and more about selectivity.

Instead of trying to understand every new tool, choose one narrow area of your life where AI might reduce friction. It might be drafting emails more quickly, summarizing long documents, organizing travel plans, helping a child outline a project, or structuring your own thinking before a meeting. Start there. Use one tool consistently for a few weeks. Notice what works and what does not. Let firsthand experience replace vague anxiety.

Instead of following every new voice claiming expertise, narrow your inputs. Choose a small number of sources that feel measured rather than breathless. Give yourself permission to ignore the rest. The goal is not to be first. The goal is to be steady.

Instead of telling yourself you need to "learn AI," ask a smaller question: what problem am I trying to solve? Then explore whether a tool can help with that specific problem. Overwhelm often comes from undefined scope. Focus begins when the scope of the problem narrows.

You do not have to move at the speed of your feed. The environment may be accelerating, but your response does not need to be frantic. Evolution does not require panic. It requires intention.

If you read this and think, "Yes, I am overwhelmed, but I am not ready to change that," that is your choice. The purpose here is not to rush you. It is to offer a path toward steadiness if and when you decide you want one.

The Overwhelmed posture is understandable in a world saturated with noise. The Focused posture is available when you are ready.

PERSONA #2: THE SKEPTIC

Not everyone responds to rapid change with anxiety. Some respond with doubt.

If you have found yourself watching the surge of enthusiasm around artificial intelligence with a steady, measured distance, this may feel familiar. You read declarations that entire professions are about to disappear and think you have heard similar warnings before. You see productivity claims that sound too clean to be true and instinctively look for what is being left out. Having lived through enough cycles of disruption, you have learned that excitement often outruns reality, and you are not inclined to chase what may settle on its own.

The Skeptic does not feel flooded. The Skeptic feels unconvinced.

And in many cases, that skepticism has been earned.

Skepticism has protected people in prior waves of change. It slows momentum when others are rushing and resists fear-driven reactions. It filters exaggerated claims from grounded evidence and demands proof before adoption. In environments where certainty is rewarded and nuance is ignored, skepticism can feel like discipline rather than resistance. It can feel like maturity formed through experience rather than enthusiasm formed through novelty.

This posture shows up in practical and often reasonable ways. A manager hears about AI integrations and assumes the benefits are overstated, deciding to wait until clearer proof emerges. A professional scrolls past tutorials and concludes that most of what is being shared is marketing rather than substance. A founder observes competitors experimenting and tells herself that her industry is built on relationships and judgment that cannot be automated. A parent assumes that the concern around new technologies will fade as earlier waves did. A student decides not to engage deeply because much of the online conversation appears inflated and repetitive.

Underlying these reactions is a quiet but powerful internal logic: we have seen this before, and we were right to wait.

There is often more beneath that logic than caution alone. The Skeptic has learned that being swept up in enthusiasm carries risk. It can cost time. It can cost credibility. It can make you look naïve. There is a subtle pride in not being easily convinced, in not being one of the first to declare something transformative. Being right about past hype cycles reinforces that instinct. It confirms that patience and restraint were wise.

That history creates a protective reflex. When bold claims surface, the reflex activates quickly. The underlying message is simple: do not be fooled.

I have felt that reflex myself. When the first wave of generative AI tools began appearing everywhere, I remember assuming that much of what I was seeing would settle as earlier cycles had. In some cases, that instinct preserved focus and prevented distraction. In others, it delayed my recognition that something more structural was unfolding beneath the surface.

What has shifted in this cycle is not the absence of hype, but the depth of integration. Artificial intelligence is no longer arriving only as standalone products that can be adopted or ignored. It is embedding itself into systems people already rely on. Email platforms incorporate it. Office software incorporates it. Search engines incorporate it. Creative tools incorporate it. Customer service platforms incorporate it. Research and analytics platforms incorporate it. Even without intentionally adopting a new AI application, you are likely interacting with AI features through systems that feel stable and familiar.

That integration changes the calculus.

In previous waves of innovation, skepticism functioned as a filter. Platforms rose and fell visibly. Letting early adopters experiment protected time and energy. Waiting allowed the noise to settle before committing. Today, integration happens quietly and incrementally. Features appear inside tools you already use. Capabilities improve in the background. Expectations adjust without formal announcements.

When skepticism turns into broad disengagement, it can create a narrowing of exposure that is difficult to see from the inside. You may underestimate how colleagues are reducing preparation time through AI-assisted research. You may not notice how expectations around drafting, analysis, or responsiveness are adjusting. You may assume that because many claims are inflated, most claims are, and in doing so miss incremental but meaningful capability.

The cost of remaining entirely skeptical is rarely immediate or dramatic. It does not usually result in obvious failure. Instead, it creates subtle but compounding friction.

You may find yourself in meetings where others reference tools you have chosen not to explore, and while you remain confident in your reasoning, you feel less fluent in the discussion. You may hesitate to pursue opportunities that reference AI familiarity because you have not tested the tools yourself. You may decline to introduce new workflows to your team because you remain unconvinced of their value. Each of these decisions feels measured in isolation. Accumulated over months, they can create distance from evolving norms.

Distance built on assumption tends to reinforce itself. When opinions are formed without direct interaction, they are rarely updated. Skepticism that once sharpened discernment can gradually harden into dismissal, not because the evidence demanded it, but because exposure never occurred.

The posture that protected you from hype can gradually insulate you from reality.

None of this suggests that skepticism should be abandoned. Skepticism remains valuable in an environment saturated with exaggeration. The evolution here is not toward enthusiasm but toward pragmatism.

Pragmatism preserves discernment while adding experience. It does not chase every release, nor does it accept bold claims at face value. What changes is the willingness to test before dismissing. Rather than rejecting broadly, the Pragmatist experiments selectively and evaluates honestly. The goal is not adoption for its own sake. The goal is informed judgment grounded in firsthand understanding.

In practice, this might mean choosing one recurring task and testing AI assistance for a defined period, not because you believe the marketing, but because you want evidence. It might involve using AI to summarize a long document and comparing the output to your own summary, not to replace your thinking but to understand its limits. It might mean asking a colleague to demonstrate exactly how they are using a tool rather than reacting to headlines. It might involve observing how integration is quietly reshaping expectations in your field, even if you remain cautious about broader implications.

Pragmatism requires humility. It acknowledges that prior experience, while valuable, may not perfectly predict current conditions. It allows new information to update old assumptions.

It does not abandon judgment; it calibrates it. It recognizes that refusing to look does not preserve objectivity. It limits it.

Experience should sharpen judgment rather than narrow it.

If you read this and feel comfortable remaining skeptical, that is your choice. Skepticism has protected you before, and it may continue to do so. The invitation is not to replace doubt with enthusiasm, but to refine doubt with experience so that it continues to serve you in an environment where change is embedding quietly and persistently rather than announcing itself loudly. In earlier cycles, skepticism filtered out noise effectively. In this one, pragmatism may prove to be the more durable expression of that same instinct.

PERSONA #3: THE EXPOSED

From Exposed to Equipped

Not everyone feels overwhelmed by the pace of change. Not everyone feels skeptical of it either. Some people feel something quieter and more personal.

They feel exposed.

Exposure does not usually arrive as fatigue the way being overwhelmed often does. It is not primarily about too much information. Nor is it a posture of doubt, as skepticism often is. It is the sense that the conversation has moved forward and you are no longer certain where you stand inside it. You may not feel flooded or unconvinced. You may simply feel slightly behind.

You recognize it in moments that pass quickly but linger privately.

You are in a meeting and someone mentions using AI to refine a proposal. The conversation moves on, and you realize you do not actually know what tool they used or how they structured the prompt. You are early in your career and sense that you are expected to be digitally fluent, yet your experience is shallow. You are mid-career and used to being competent in most rooms, yet this topic makes you cautious. You are senior and expected to shape direction, but you have not spent enough time hands-on to feel grounded in specifics.

At home, exposure looks different but feels similar. Your child references using AI to outline a paper. You nod, but you are unsure what that process actually looks like. A colleague shares a polished output and you suspect assistance was involved, though you are not certain how much. A friend speaks confidently about how these tools are reshaping work, and you quietly wonder whether you have underestimated their impact.

In each case, the discomfort is not loud. It is the subtle fear of being found out.

This posture makes sense.

Competence is tied to identity. Many of us have built careers on being capable, informed, and steady. When a new layer of capability spreads quickly across industries and age groups, it can create the impression that prior competence is being re-scored. Even if our core skills remain valuable, we may feel as though the baseline has shifted.

There is also a social distortion at work. Online, people rarely share their learning curve. They share outcomes. They present finished drafts, optimized workflows, dramatic time savings. When we compare our private uncertainty to someone else's curated confidence, exposure intensifies. It becomes easy to assume that others are further ahead than they are.

Exposure differs from overwhelm in an important way. The overwhelmed person often avoids engagement because there is too much coming at them. The exposed person avoids engagement because they do not want to confirm a gap. One is saturated. The other is self-conscious.

It also differs from skepticism. The skeptic holds back because they are unconvinced. The exposed person often suspects there is real value here, but hesitates because they feel underprepared to engage credibly.

Avoidance becomes the quiet coping strategy. You skim headlines but do not test tools. You defer exploration to others. You contribute when discussions remain abstract, but retreat when they become practical. You tell yourself that your experience still matters, which is true, yet you leave a small gap unaddressed.

Over time, that gap grows larger in your mind than it is in reality.

The cost of staying exposed is rarely dramatic. It does not usually result in failure. Instead, it shows up as hesitation. You hesitate to speak when AI is mentioned. You hesitate to pursue opportunities that reference AI familiarity. You hesitate to guide others because you are unsure how deep your own understanding runs. Each hesitation is small. Accumulated, they narrow your presence in evolving conversations.

There is another cost as well. When you rely exclusively on secondhand commentary, your perception of the landscape is shaped by other people's interpretations. Some of those interpretations are thoughtful. Others are inflated. Without firsthand interaction, it is difficult to tell what is genuinely useful and what simply looks impressive.

The evolution here is not toward mastery. It is toward being equipped.

Equipped does not mean becoming the most knowledgeable person in the room. It means having enough direct experience to participate without insecurity. It means being able to say, calmly and honestly, "I have tried this. Here is what I found useful. Here is where it falls short." It replaces imagined deficiency with grounded familiarity.

Moving from exposed to equipped does not require immersion. In fact, trying to immerse yourself completely may push you toward overwhelm. The shift is narrower than that.

If you are early in your career, choose one area where AI can support your development without replacing your effort. Use it to structure research, then refine the analysis yourself. Ask it to critique a draft before submission. Treat it as a supplement to your thinking rather than a substitute. The goal is to understand how it behaves, not to lean on it reflexively.

If you manage others, experiment visibly and modestly. Share what you are testing and what you are learning. Let your team see that you are closing the gap deliberately rather than pretending it does not exist. Leadership credibility does not require instant fluency. It requires honest engagement.

If you operate at a senior level, allocate time for direct interaction rather than relying entirely on briefings. Draft something with assistance and compare it to your own version. Ask a model to summarize a strategic issue and examine where nuance is lost. Your judgment remains the differentiator. Hands-on exposure sharpens it.

If you are a teacher, coach, or parent, explore the tools your students or children are already using. Run an assignment prompt yourself and see what emerges. Ask them to explain their process. Shared exploration often reduces the generational tension that distance amplifies.

In each case, limit the scope. Select two or three tools at most. Use them repeatedly over several weeks. Familiarity builds through repetition, not through sampling everything once. The objective is not to keep up with every release. It is to replace speculation with experience.

You may discover that the gap was smaller than you imagined.

Equipped people do not know everything. They know enough. Enough to ask better questions. Enough to contribute. Enough to decide where deeper investment is warranted and where it is not.

If you recognize yourself here and feel no urgency to adjust, that is your choice. Exposure does not invalidate your existing competence. It simply highlights an area where modest engagement could restore steadiness. You do not need to transform your identity. You need to reduce the distance between what you suspect and what you have actually tried.

The goal is not to become the expert.

It is to stand comfortably in the room.

From that position, confidence tends to rebuild naturally.

PERSONA #4: THE SEEKER

From Seeker to Employed

Some people are not trying to optimize their workflow or debate what artificial intelligence means for the future. They are trying to get in, or they are trying to get back in.

If you are sending out applications and hearing very little in return, this chapter may feel close to home. If you were laid off from a stable role and are now re-entering a market that feels tighter and more technical than when you left, this chapter may feel equally familiar. If you are early in your career and noticing that entry-level postings seem fewer, more demanding, or strangely compressed, you are not imagining it.

You are not overwhelmed by headlines. You are not skeptical from a distance. You are not worried about sounding uninformed in a meeting. You are standing outside the gate, or standing just inside it, trying to secure footing that feels stable.

The anxiety is practical.

You see tools generating research summaries, draft proposals, slide decks, marketing copy, and financial models in seconds. You notice that tasks which once trained beginners — organizing information, preparing first drafts, assembling reports — are now assisted by software before a junior employee ever touches them. You read job descriptions that assume familiarity with AI tools as a baseline rather than a bonus. You hear that screening systems filter résumés long before a human decision-maker reviews them.

It is not unreasonable to wonder whether the ladder has narrowed.

This concern is not limited to recent graduates. Mid-career professionals who lose roles in restructuring cycles often feel it even more sharply. They may have built years of experience on work that involved drafting reports, analyzing data, preparing presentations, or coordinating information across teams. When they re-enter the market, they discover that much of that work can now be accelerated by software. Their competence has not vanished, but the way it is valued may have shifted.

Work itself is not disappearing wholesale, but certain layers of it are being compressed. In earlier waves of automation, disruption was most visible in factories and other physically demanding industries, where machines replaced repetitive manual tasks first. In this cycle, the early compression is more visible in jobs built around documents, data, reports, emails, and screen-based analysis. Tasks that once filled the first months or years of a career can now be completed in minutes with assistance.

Employers notice this shift. They may not eliminate roles entirely, but they often consolidate responsibilities or expect broader capability from each hire. Opportunity still exists, but what is fading is generic entry.

In earlier cycles, effort and willingness to learn on the job were often enough to secure a foothold. You could begin with repetitive tasks, build competence gradually, and move upward over time. Today, some of that repetition is automated. The visible work skews more toward interpretation, integration, and judgment from the outset.

If you approach the market competing primarily on effort, formatting, or surface credentials, the process can feel longer and more uncertain. The challenge is not a lack of ability. It is that the way people are compared has shifted slightly upstream.

Remaining purely a Seeker is not wrong. Many people eventually find employment through persistence, relationships, and timing. Markets still hire. Organizations still need people. However, relying only on patience in a compressed environment can extend instability and increase vulnerability once hired.

And instability carries weight.

For those who have already experienced displacement, the fear is rarely limited to getting hired. It extends to staying hired. Losing a role once can shake confidence and financial stability. Losing it twice in quick succession can alter how you think about risk, career choices, and long-term planning. Even for those early in their careers, entering a role that feels fragile can create constant low-level tension.

Employment in this moment is not only about income. It is about durability.

The evolution here is toward becoming employed in a durable sense. In other words, it is the shift from being the Seeker to becoming one of the Employed — someone whose contribution allows them to remain inside the organization as conditions continue to evolve rather than constantly standing outside the gate hoping it opens again.

Being employed in this sense does not mean simply having a job. It means contributing in a way that reduces replaceability. It means understanding how value is being recalibrated and adjusting before that recalibration forces the adjustment upon you.

What does that look like?

First, it requires clarity about what software now handles easily. If a tool can generate a competent first draft, summarize a document, or organize information quickly, then competing on first drafts alone is unlikely to distinguish you. The differentiator shifts to how you refine, question, and connect that output to context. When you demonstrate the ability to interpret rather than merely produce, you begin to resemble the kind of person who remains employed even as tools improve.

For a recent graduate, this may mean showing not just completed assignments but the reasoning behind your choices. If you used AI to assist with research, be prepared to explain how you evaluated its suggestions and where you adjusted them. Employers increasingly assume baseline tool usage, but what they examine more closely is judgment.

For someone re-entering the market, this may mean reframing prior roles in terms of decisions influenced and outcomes shaped rather than tasks completed. If much of your work involved producing reports, describe how those reports informed strategy or changed behavior. Remaining employed over time grows from visible impact rather than the sheer volume of work produced.

Second, becoming employed in a durable sense requires familiarity with modern tools without allowing those tools to become your identity. Organizations are not looking for hype. They are looking for contributors who can operate in environments where automation is present. Being able to describe calmly and specifically how you integrate tools into your process signals adaptability. Overstating expertise often signals insecurity, while measured familiarity signals steadiness.

Third, it requires thinking slightly upstream from the tasks most easily automated. If routine analysis is accelerated by software, human contribution shifts toward framing better questions, identifying trade-offs, and making decisions under ambiguity.

These skills can be developed before employment. Small case studies, personal projects, volunteer initiatives, or freelance work can demonstrate how you approach unclear problems. A modest portfolio that reveals how you think often communicates more than a résumé optimized only for keywords.

The urgency in this persona is real because the consequences are tangible. A prolonged search affects finances and confidence. Fragile employment amplifies stress. At the same time, panic is not productive. Competing directly with machines on speed or volume rarely creates stability. Positioning yourself as someone who exercises judgment in a machine-assisted environment is far more likely to do so.

You can remain a Seeker and trust that persistence will eventually open a door. That path remains available. The invitation here is to improve both the likelihood of entry and the durability of what follows by adjusting how you define and demonstrate value.

Getting in matters.

Remaining employed matters just as much, because the deeper shift in this chapter is the movement from Seeker to the Employed — the people whose contribution makes them harder to remove when conditions tighten.

The gate has not disappeared. It has shifted slightly upward, and it tends to open more readily for those who arrive prepared to contribute beyond what automation now handles alone.

PERSONA #5: THE CURIOUS

From Curious to Capable

You have probably signed up for more AI tools than you actively use. You have downloaded apps, opened trial accounts, experimented with image generators, rewritten paragraphs through different models, compared outputs, and bookmarked tutorials that promised to unlock new efficiencies. Each new release feels like an opportunity. Each experiment carries a small surge of possibility.

You are not resistant to change. If anything, you lean toward it. When something new appears, you want to understand it. You test features quickly. You explore integrations. You enjoy seeing what is possible. You may even be the person others turn to when they want to know what a new tool can do.

Curiosity is not a weakness. In many ways, it is a strength.

Curiosity keeps you from becoming rigid. It keeps you open to improvement. It allows you to imagine different workflows, new creative paths, and alternative approaches to old problems. While others hesitate, you explore. While others debate, you try.

Over time, however, a pattern can form.

You subscribe to one tool and then another. You purchase a course that promises to help you master prompting. You test a productivity assistant for a few weeks and then shift to a different one that seems more advanced. You experiment with automating your schedule, organizing your notes, improving your writing, generating ideas for a side project. The surface area of your experimentation expands steadily.

I recognize this pattern in myself. There have been stretches where I signed up for tools faster than I learned them, assuming that exposure alone meant progress. It took deliberate restraint to realize that sampling widely and integrating deeply are not the same thing.

Outside of formal work, this can look like using AI to plan travel, draft creative writing, design personal branding, outline fitness routines, manage household logistics, or structure learning plans. You enjoy seeing how quickly a system can respond. You appreciate the speed and novelty. You often feel up to date.

Inside work, curiosity may show up as suggesting new tools, volunteering to test integrations, or demonstrating features to colleagues. You are rarely behind the curve. You are usually aware of what is emerging.

Yet after months of experimentation, something subtle becomes apparent. You have touched many tools, but you rely deeply on very few. You know what is possible in theory, but you have not built sustained competence with most of what you have tried. When faced with a demanding project, you often default to older habits because the newer tools never fully integrated into your routine.

Curiosity creates exposure. It does not automatically create capability.

This posture makes sense in the current environment. New tools appear constantly. Features update weekly. Entire categories seem to evolve in real time. The culture surrounding AI rewards early discovery and visible experimentation. Being the first to test something often feels like progress.

There is also a psychological reward in novelty itself. Each new tool offers anticipation. You imagine how it might improve your output or simplify your process. That anticipation can feel productive, even before anything concrete has changed. The act of exploring becomes a form of engagement that signals adaptation.

The difficulty is that novelty competes with depth.

When every week introduces something new, sustained practice becomes harder. Instead of refining your use of one platform, you divide your attention across many. Instead of learning how to evaluate output more critically, you move to the next system that promises better results. Instead of building familiarity through repetition, you move on before that familiarity has time to develop.

Unlike the Seeker, you are not primarily concerned with securing employment. Unlike the Exposed, you are not avoiding engagement. Unlike the Skeptic, you are not holding back judgment. You are participating actively and willingly.

The risk is not hesitation but diffusion.

Diffusion has a quiet cost. Subscriptions accumulate. Courses stack. You invest time setting up workflows that you abandon when something more interesting appears. Because you are consistently exploring, it feels as though you are evolving. In reality, your progress spreads outward rather than deepening.

Over time, this creates an illusion of fluency. You can speak about many tools. You recognize their interfaces. You understand their general capabilities. Yet when a complex task demands sustained interaction, you may feel less confident than expected because you have not committed long enough to any one system to understand its nuances.

Curiosity without direction can become a loop.

The evolution here is toward becoming Capable.

The Capable are not the people who try every tool first. They are the people who choose carefully and stay with their choices long enough to develop real competence. Moving from Curious to Capable often requires becoming comfortable with not always being the first person out of the gate.

Early experimentation can be valuable, but capability tends to grow when novelty slows down long enough for familiarity to develop. The Capable understand that depth usually matters more than speed.

Capable engagement does not mean mastering every new release or staying ahead of every update. It means selecting a small number of tools that align with your needs and developing real familiarity with them. It means understanding not only what they can do, but where they fall short. It means integrating them into your routines rather than testing them sporadically.

Capable engagement often looks quieter than curiosity.

It may not generate the same visible excitement. It involves repetition and sustained interaction. It involves returning to the same interface long enough to learn its edges. It involves resisting the urge to replace a tool the moment something newer appears.

In personal life, this might mean choosing one assistant for planning and using it consistently instead of rotating among several. It might mean committing to drafting through one platform until you understand how to refine its suggestions effectively. It might mean finishing a course before purchasing another, allowing the learning to consolidate rather than fragment.

In work settings, it may mean narrowing experimentation and focusing on building repeatable processes. Instead of testing every integration, you identify where automation genuinely improves quality or efficiency and refine that process over time. Familiarity builds confidence, and confidence builds speed.

Depth creates compounding returns. When you use the same tools consistently, you begin to notice patterns. You understand how to prompt more precisely. You recognize where the system is likely to misinterpret context. You learn how to combine features in ways that feel natural rather than forced. Over months, this depth becomes visible in the quality and reliability of what you produce.

There is also a financial and cognitive dimension. Endless sampling can quietly drain both money and attention. Subscriptions renew. Time fragments. The sense of being perpetually in beta mode can prevent satisfaction with what you already have. The Capable tend to invest where they commit and ignore what they are not ready to integrate.

Remaining Curious is not inherently harmful. Curiosity often sparks discovery and creativity. The invitation in this chapter is simply to notice whether curiosity is building lasting capability or merely sustaining stimulation.

You do not need to stop exploring. You may only need to narrow your focus long enough for competence to deepen.

Curiosity often opens doors, but the Capable build something behind those doors that lasts.

PERSONA #6: THE OUTDATED

From Outdated to Relevant

There is a particular kind of discomfort that does not feel like fear and does not feel like confusion. It feels more like distance.

You find yourself in conversations where artificial intelligence is referenced casually, as if everyone has already experimented with it. A family member mentions using it to plan a trip or organize a school project. Someone jokes about a video that turned out to be generated by software. A headline scrolls past with terms that assume familiarity. You understand the broad idea, but you sense that the center of the conversation has shifted slightly beyond your daily habits.

The feeling is subtle. It resembles realizing you no longer recognize the names of the bands everyone else seems to know. You may not want to follow every new artist or memorize every lyric, but there is a moment when you notice that references are moving more quickly than your awareness. You smile, nod, and let the conversation continue, aware that you are listening more than participating.

This has happened before.

Every generation eventually reaches a point where new language, new tools, and new cultural references spread faster than their routines adjust. What once felt intuitive becomes current for someone else. What once signaled fluency becomes background. Artificial intelligence is simply the newest expression of a long-standing pattern in which culture renews itself and familiarity shifts.

Feeling outdated does not mean you are incapable.

You may have navigated earlier technological transitions. You may have adapted to email, mobile phones, social media, and countless other changes. You may carry experience and judgment that only time can build. The discomfort does not come from a lack of intelligence. It comes from not being immersed in the spaces where the newest references circulate.

Unlike the Exposed persona, this is not about fearing a specific moment of embarrassment. The Exposed worry about being caught unprepared in a meeting or task. The Outdated experience something broader and slower. It is not a single event. It is a gradual widening of space between your habits and the cultural edge.

Unlike the Skeptic, this posture is not rooted in doubt. You may see the value of these tools. You may even be intrigued by them. You simply have not spent enough time near them to feel comfortable. And unlike the Curious, you are not experimenting widely. You are observing.

That observation can quietly become distance.

At first, the distance is minor. You let others explain new terms. You allow conversations to move forward without asking clarifying questions. You assume that these developments are more relevant to younger people and that you can afford to remain slightly removed. Over time, however, small removals accumulate.

You may begin to speak less when certain topics arise. You may defer to others when technology intersects with broader discussions. You may find yourself translating references internally rather than engaging with them directly. The more you observe from the edge, the more natural it feels to remain there.

That shift is rarely visible from the outside. It happens internally. You begin to see yourself as slightly less current and slightly less connected to what feels contemporary. Not incapable. Not irrelevant. Just a step removed.

The cost of remaining here is not dramatic, but it is real.

Cultural participation affects confidence. When you feel close to what is happening, you speak with ease. When you feel distant, you hesitate. Over time, hesitation can narrow your sense of voice. You may unconsciously limit your contribution in conversations that blend technology with ethics, culture, or society, even though your perspective is shaped by years of lived experience.

Relevance in this context is not about mastery.

It is about proximity.

The evolution here is toward becoming Relevant, not by chasing every development but by staying close enough to understand the conversation as it unfolds. The Relevant are not the people who know every tool or follow every trend. They are the people who remain near enough to the cultural edge that they can participate without feeling lost.

You do not need to understand every technical detail. You do not need to adopt every new habit. You do not need to immerse yourself in online communities that move at a pace that does not suit you. The shift from Outdated to Relevant is modest but meaningful. It involves stepping close enough that you remain part of the conversation rather than watching it from a distance.

Relevant does not mean fluent in every feature. It means comfortable enough to recognize references, ask thoughtful questions, and form opinions grounded in light firsthand experience.

That shift can begin in simple ways.

You might choose one practical use case and explore it personally. Perhaps you use AI to summarize an article you intend to read, draft a message, generate ideas for a hobby, or organize information you would otherwise structure manually. The purpose is not efficiency alone but familiarity. Direct interaction replaces abstraction.

If younger family members or friends use these tools frequently, you might ask them to show you how they incorporate them into daily life. Not as a test and not as a critique, but as shared exploration. When curiosity replaces distance, generational gaps often narrow quickly. You may not adopt every habit, but you gain context, and context restores comfort.

You might also choose to stay lightly informed through steady, balanced sources rather than constant feeds. Reading an occasional thoughtful explanation or watching a clear demonstration can provide enough grounding to follow conversations without strain. A modest amount of exposure often closes more distance than endless observation.

The goal is not to become immersed in the latest release.

The goal is to remain near enough to cultural shifts that you retain your voice.

There is dignity in experience. There is also dignity in adaptation. Experience provides perspective that newer participants may not yet have. Adaptation ensures that perspective continues to matter in contemporary spaces. The Relevant hold both at once. They do not abandon their experience, but they also do not allow distance to grow so wide that their perspective loses its place in the conversation.

Remaining Outdated is not a failure. Many people choose to engage selectively with new developments and live fully connected lives. The invitation here is simply to notice whether distance is beginning to shape how you see yourself or how willing you are to participate in conversations that intersect with the present moment.

You do not need to follow every new band or understand every lyric.

You may only need to recognize enough of the melody that you can join in when it matters.

Relevance is not reinvention.

It is staying close enough that your experience continues to meet the present rather than sit beside it.

SO NOW WHAT'S NEXT FOR YOU?

By now you have likely seen yourself somewhere in these pages. You may have recognized the fatigue of being overwhelmed, the restraint of skepticism, the quiet discomfort of feeling exposed, the pressure of trying to get in or get back in, the restless experimentation of curiosity, or the subtle distance that can develop over time. Most of us do not remain in only one of these postures. We move between them depending on what the moment demands.

The question that opened this book was simple: So now what?

Up to this point, that question has been reflective. You have been observing your own habits and reactions. You may have paused at certain passages and thought, "That is accurate." Recognition brings clarity. It reduces the vague tension that comes from not quite understanding why something feels off. It helps name patterns that otherwise operate quietly in the background.

Clarity, however, does not automatically lead to change.

There is a familiar rhythm that often follows insight. You close the book with the sense that it was helpful. You carry a few ideas with you into the rest of your day. You may even intend to adjust something. Then your normal routines resume. The same inputs return. The same habits fill the available space. The small shift you imagined making becomes one more good intention waiting for a better moment.

I have repeated that rhythm more times than I would like to admit. I have purchased books that described exactly what I needed to see and left them neatly stacked beside my desk. I have enrolled in courses that promised focus and completed the first section with energy, only to let the rest sit untouched once the initial momentum faded. I have listened to thoughtful talks that sharpened my thinking and then continued operating as before because continuing as before felt easier.

Agreement is comfortable, but turning that agreement into action requires friction.

There are readers who will finish this book and recognize that they have not begun in any deliberate way. They have followed the conversation about artificial intelligence from a distance, postponing direct engagement and telling themselves they will step in when they have more time or more certainty. For them, the next step does not need to be large. It simply needs to be real. Choosing one concrete action interrupts the pattern of waiting.

There are also readers who have already begun. They experiment with tools. They use AI to curate music, summarize articles, draft messages, or organize tasks. They are not standing still. Their challenge is different. It lies in deciding whether what they are doing is intentional or incidental. Light experimentation can create the sense of progress without producing meaningful change in posture.

Starting matters. So does deepening.

The reason to move, whether you are beginning or refining, is not performance. It is stability. When you act deliberately, even in small ways, the environment feels less abstract. Testing something yourself replaces speculation with experience. Narrowing one source of noise makes attention steadier. Closing one small gap reduces the quiet tension that comes from feeling slightly behind. These adjustments may seem modest, but over time they shape how confident and present you feel in conversations that continue to evolve.

If overwhelm resonated, narrowing one input protects your attention rather than scattering it further. If skepticism felt familiar, testing one assumption strengthens judgment rather than weakening it. If exposure described you, spending time with one tool replaces uncertainty with familiarity. If you are seeking entry or re-entry, refining how you demonstrate your thinking increases durability rather than simply adding effort. If curiosity defines your habits, committing to depth builds skill instead of spreading it thin. If you felt slightly outdated, remaining close to one conversation keeps you connected rather than observing from the edge.

You do not need to do all of these things. You need to do one of them.

Before you set this book aside, decide what will change in your behavior over the next few days because you read it. Choose something specific enough that it cannot dissolve into abstraction. It might be an hour spent exploring a tool you have avoided. It might be canceling a subscription that crowds your attention. It might be revising a paragraph to better reflect your reasoning. It might be asking someone to show you how they use a system you have only heard about. The action itself is less important than the fact that it is chosen.

If you are already moving, examine your movement honestly. Are you shaping it deliberately, or are you participating casually because participation is easy? Are you deciding where to focus, or following whatever appears next? Refining your engagement can be as meaningful as initiating it.

If you want evidence that this approach is more than theory, return to how this book came to be.

I could have postponed writing it. I could have told myself that I needed more research, more time, or a quieter stretch in the calendar. I could have outlined these ideas and returned to them months later when the pace felt less intense. Instead, I chose to use the tools available to me to move the idea forward. The ideas are mine. The experiences are mine. The judgment about what belongs in these pages is mine. What changed was the decision to act rather than wait for ideal conditions.

As a result, this book was drafted, refined, and published in seven days.

The point is not the speed but the decision not to delay. Waiting would have preserved comfort. Acting required a small push beyond it.

You do not need to write a book in a week. You do not need to compress your timeline dramatically. It is enough to notice where you are postponing a small adjustment and to make it now rather than later.

"So now what?" does not disappear once answered. It returns as circumstances shift. You will likely feel overwhelmed again, or skeptical, or curious, or slightly behind. That is part of living in a period of rapid change. What matters is whether you recognize those moments sooner and respond with intention rather than habit.

The tools will continue to evolve, the pace will continue, and the conversation will move forward.

What remains steady is your ability to choose how you engage with it.

So now what's next for you?

Not in theory but in practice, and not someday but this week.

Choose one adjustment that aligns your behavior with what you already know is worth doing. Take that step. Let it change how you feel in the next conversation, the next task, the next quiet moment when you would otherwise drift.

From there, the next step will make itself visible.

That is enough.

ABOUT THE AUTHOR

Tom Mawhinney has spent more than three decades observing how organizations and individuals respond when the world around them begins to change. His career has taken him through boardrooms, executive teams, entrepreneurial ventures, and advisory roles across multiple industries and economic cycles.

Along the way, he developed a particular interest in how people interpret and adapt to technological disruption. From early waves of digital transformation to today's rapid advances in artificial intelligence, Tom has worked closely with organizations trying to understand what these shifts mean for their people and their future.

Through his writing and speaking, he explores the human side of change—how individuals make sense of disruption, how organizations evolve, and how thoughtful decision-making can turn uncertainty into opportunity.

So Now What? reflects that perspective. Rather than predicting where artificial intelligence is headed next, the book focuses on how people are responding to the moment we are already living through.

Email: tom@tommawhinney.io

Website: www.tommawhinney.io

LinkedIn: https://www.linkedin.com/in/tommawhinney

NOTES

NOTES

NOTES

NOTES

NOTES

NOTES

NOTES

NOTES

NOTES

NOTES

9 781968 318567